THE POETRY OF COPPER

The Poetry of Copper

Walter the Educator™

Silent King Books

Copyright © 2023 by Walter the Educator™

All rights reserved. No part of this book may be reproduced in any manner whatsoever without written permission except in the case of brief quotations embodied in critical articles and reviews.

First Printing, 2023

Disclaimer
This book is a literary work; poems are not about specific persons, locations, situations, and/or circumstances unless mentioned in a historical context. This book is for entertainment and informational purposes only. The author and publisher offer this information without warranties expressed or implied. No matter the grounds, neither the author nor the publisher will be accountable for any losses, injuries, or other damages caused by the reader's use of this book. The use of this book acknowledges an understanding and acceptance of this disclaimer.

"Earning a degree in chemistry changed my life!"
- Walter the Educator

dedicated to all the chemistry lovers, like myself, across the world

CONTENTS

Dedication V

Why I Created This Book? 1

One - Oh, Copper 3

Two - Copper's Touch 5

Three - Conductor Of Stories 7

Four - Copper, Forever 9

Five - Eternal Connection 11

Six - Beauty And Might 13

Seven - Celestial Sphere 15

Eight - Timeless Reflection 17

Nine - Balance And Peace 19

Ten - Greenish Hues 21

Eleven - Metal That Sparkles 23

Twelve - Adorned The Crown 25

Thirteen - Reddish Hue 27

Fourteen - Warmth And Grace 29

Fifteen - Forged In The Core 31

Sixteen - Pennies That Glimmer 33

Seventeen - Ages Past 35

Eighteen - Earthly Space 37

Nineteen - Craftsmanship's Embrace 39

Twenty - Silent Hero 41

Twenty-One - Sight To View 43

Twenty-Two - Copper Is Versatile 45

Twenty-Three - Wonder And Grace 47

Twenty-Four - Ancient Egypt 49

Twenty-Five - Precious Gift 51

Twenty-Six - Unique And Rife 53

Twenty-Seven - Copper, A Muse 55

Twenty-Eight - Metal Divine 57

Twenty-Nine - Breaking Rules 59

Thirty - Strength And Durability 61

Thirty-One - Unbroken And Raw 63

Thirty-Two - Conduct Heat And Electricity 65

Thirty-Three - Magic In The Sand 67

Thirty-Four - Ignites The Dark 69

Thirty-Five - Metal Of Might 71

About The Author 73

WHY I CREATED THIS BOOK?

Creating a poetry book about the chemical element of Copper was a unique and captivating endeavor for several reasons. Firstly, Copper is a versatile and essential element with a rich history and various interesting properties. Exploring its characteristics, symbolism, and applications through poetry can offer a fresh perspective on this element.

Poetry has the power to evoke emotions, capture imaginations, and convey complex ideas in a concise and artistic manner. By crafting poems centered around Copper, I can delve into its physical attributes, such as its reddish-brown hue, malleability, and conductivity. Additionally, I can explore its historical significance, cultural associations, and metaphorical representations.

Furthermore, Copper has connections to human experiences, relationships, and emotions. It has been used in ancient artifacts, architectural designs, and artistic creations. It is also an essential element in our daily lives, found in electrical wiring, plumbing systems, and even our bodies. By incorporating these aspects into poetry, I can create a captivating collection that intertwines science, history, and personal reflections.

This poetry book about Copper can also serve as a metaphorical exploration of human qualities and experiences. Copper is known for its durability, its ability to conduct energy, and its ability to change and evolve over time due to oxidation. These characteristics can be used symbolically to explore themes such as resilience, transformation, and the interconnectedness of people and nature.

Overall, creating this poetry book about Copper offers a unique opportunity to blend science, history, and personal experiences in a way that is both educational and emotionally resonant. It allows me to explore the beauty and significance of an element often taken for granted and present it in a new and thought-provoking light.

ONE

OH, COPPER

In the depths of Earth's embrace,
A treasure lies beneath the space,
A metal born with lustrous grace,
Copper, the element we chase.

From ancient times, its tale was told,
A metal precious, a sight to behold,
With hues that shimmer, red and gold,
Copper's allure, a story to unfold.

Born in the fiery heart of stars,
Forged in cosmic furnaces afar,
Copper, the conductor of electric charge,
Unveils its secrets, a celestial mirage.

A gift to artisans, a sculptor's dream,
Malleable and soft, it takes the gleam,

Crafted into vessels, a brewer's theme,
Copper, the alchemist's timeless scheme.

    In wires, it hums with vibrant might,
Guiding currents through the darkest night,
A conductor of energy, a beacon of light,
Copper, the conductor, shining bright.

    In ancient lore, a symbol of love,
A penny's worth, a talisman above,
Copper coins, exchanged with a gentle glove,
Copper, the symbol, of connections thereof.

    From rooftops, gleaming under the sun,
Weathered over time, its journey begun,
Copper, the protector, when storms come undone,
A guardian of shelter, forever one.

    Oh, Copper, element of allure,
Your story timeless, forever pure,
In your beauty, we stand secure,
Copper, the element we shall endure.

TWO

COPPER'S TOUCH

In the realm of nature's wonders, behold the element so bold,
A metal that gleams with an ancient story untold.
Copper, a gift from the Earth's secret vaults,
A treasure that beckons and never halts.

Amidst the darkness of Earth's deep embrace,
Copper emerges with its fiery grace.
A conductor of electricity, it dances with might,
Connecting the realms of day and night.

In copper's hues, a symphony of shades,
From fiery orange to earthy braids.
Its surface, weathered by time's gentle touch,
A testament to resilience, it speaks so much.

From ancient civilizations to modern feats,
Copper's presence in history truly completes.

Adorning statues, roofs, and pots,
It weaves tales of strength and connects the dots.
   With healing properties, it soothes the soul,
A balm for ailments, making us whole.
Through copper's touch, a balance is found,
In body and mind, harmonies resound.
   So, let us cherish this metal so divine,
With its lustrous glow, a beauty that's fine.
Copper, a symbol of strength and grace,
Forever embedded in time's eternal embrace.

THREE

CONDUCTOR OF STORIES

In the depths of the earth, where magma flows,
Lies a treasure hidden, a secret that glows.
A metal so noble, with a lustrous sheen,
Copper, the element, in nature's serene.

Born in the stars, forged in cosmic fire,
Copper dances with passion, a burning desire.
Its atoms, like dancers, move in a trance,
Weaving a tapestry of copper's advance.

A conductor of energy, it conducts with ease,
Electricity's partner, a symphony to appease.
From wires to circuits, it breathes life anew,
Powering our world, connecting me and you.

But beyond its function, copper holds more,
A richness of history, a tale to explore.

From ancient civilizations to modern art,
Its beauty and versatility capture the heart.
In statues and sculptures, copper takes form,
Crafted by artisans, a testament to norms.
Its reddish hue, a vibrant embrace,
Transforming spaces, adding grace.
From roofs to coins, copper becomes wealth,
A currency of value, a symbol of stealth.
Its durability timeless, resisting decay,
Enduring the ages, come what may.
Copper, a metal that weaves tales untold,
A conductor of stories, precious as gold.
With every touch, its magic unfolds,
A symphony of beauty, forever to behold.

FOUR

COPPER, FOREVER

In the depths of Earth's embrace,
There lies a metal, full of grace.
Copper, radiant in its hue,
A shimmering tale, ancient and true.

Born from fiery volcanic womb,
Forged in the crucible of nature's loom.
Its atoms dance, in perfect rhyme,
Crafting a tale, through the passage of time.

A conductor of warmth, a bringer of light,
Copper weaves its magic, day and night.
In wires and cables, it conducts the flow,
Powering the world, with an electric glow.

But beyond its utility, lies a secret untold,
Legends and myths, that the ancients behold.
A metal of healing, with mystical might,
Copper's touch, soothes souls in the night.

From ancient civilizations, to the modern age,
Copper's legacy, an eternal stage.
A symbol of wealth, of wisdom and might,
It shines in the sun, reflecting life's light.

O Copper, noble element divine,
In your presence, our hearts entwine.
A testament to nature's alchemy,
Copper, forever, my muse you'll be.

FIVE

ETERNAL CONNECTION

In the depths of the earth, where secrets lie,
There dwells a metal, with a gleaming eye.
Copper, the element, so ancient and wise,
With its fiery hue, it mesmerizes.

In nature's alchemy, a gift so rare,
A conductor of energy, beyond compare.
From wires that hum with electric might,
To coins that gleam in the pale moonlight.

It adorns the rooftops, weathered and green,
Aged by time, a story unseen.
Patina of life, etched upon its face,
Whispering tales of history and grace.

In the artist's hands, a medium of art,
A canvas for passion, a delicate part.

Sculpted in beauty, forged with love,
A testament to creativity from above.
    But beneath its allure, a strength untold,
A warrior's armor, steadfast and bold.
A shield against darkness, a sword in the fray,
Copper, the element, that never shall sway.
    So let us embrace this metal divine,
With its lustrous glow, forever to shine.
For in its essence, we find a reflection,
Of resilience, beauty, and eternal connection.

SIX

BEAUTY AND MIGHT

In the heart of the earth, where secrets lie,
There dances a metal, both noble and shy.
Copper, they named it, with its fiery hue,
A gift from the heavens, forever true.

From ancient times, its allure was known,
A conductor of energy, it has shown.
In wires and cables, its power does flow,
Uniting the world with its electric glow.

A symbol of beauty, copper's artistry,
Adorns our homes with its warm patina sea.
In statues and trinkets, it tells stories untold,
Of craftsmanship and passion, as it unfolds.

But copper's tale is not just of splendor,
For in its presence, healing it renders.

From ancient remedies to modern cure,
It soothes the body, bringing wellness pure.

And deep in the ocean, where life thrives,
Copper dwells in abundance, where it dives.
A guardian of ecosystems, it fights the foe,
Protecting the balance, as only it knows.

Oh, copper, noble metal, so rare and dear,
Your legacy stretches, both far and near.
In science and art, in beauty and might,
You shine in our world, forever bright.

So let us cherish this element divine,
A symbol of strength, a treasure to find.
For copper, in all its forms, will endure,
A testament to nature's grace, so pure.

SEVEN

CELESTIAL SPHERE

In a realm of molten ores and fiery glow,
Where elements dance and secrets bestow,
There lies a metal shining bright and bold,
A tale untold, a story yet to unfold.

Copper, the alchemist's precious find,
With hues that enchant and powers bind,
A conductor of energy, gleaming so fine,
A wonder of nature's design.

In ancient realms, it was revered,
A gift from gods, it appeared,
Crafted into vessels, statues divine,
A symbol of wealth, a treasure to find.

Its touch, a remedy for the weary soul,
A healer of ailments, making us whole,

With a touch of warmth, it eases despair,
Copper, a balm for hearts in repair.

From the depths of the Earth, it does rise,
A metal that holds secrets, hidden in guise,
A conductor of dreams, a bridge to the skies,
Copper, a marvel that never denies.

So let us celebrate its beauty and grace,
In every trace, in every embrace,
For in this element, we find a connection,
To the wonders of nature's affection.

Copper, a gift from the celestial sphere,
A testament to the magic we hold dear,
In its presence, we are forever inspired,
By the alchemy of life, forever admired.

EIGHT

TIMELESS REFLECTION

In a realm of hidden depths, where dreams unfold,
Lies a tapestry of secrets, shimmering and bold.
A metal born of Earth's ancient core,
Copper, a treasure forevermore.

With hues of russet, a coppery sheen,
It gleams with stories yet to be seen.
A conductor of energy, both fierce and bright,
Dancing with electrons, an ethereal light.

From ancient times, it weaves its tale,
A metal of artisans, both humble and regal.
In every stroke of brush and chisel's mark,
Copper breathes life, leaving its mark.

In pipes and wires, it carries the flame,
Connecting worlds, igniting the same.

A conduit of power, whispers in the dark,
Copper binds us, an invisible spark.

In nature's embrace, it finds its home,
Adorning the earth, where it freely roams.
From verdant hills to canyons deep,
Copper's touch, a treasure to keep.

As we delve into its alchemical embrace,
Copper reveals its secrets, a cosmic grace.
An element of strength, resilience, and grace,
Forever entwined in this cosmic space.

So let us honor this metal so dear,
With every breath, let its essence appear.
For in copper's embrace, we find connection,
A symbol of unity, a timeless reflection.

NINE

BALANCE AND PEACE

In the realm of elements, a treasure lies,
A metal that gleams, beneath the skies.
Copper, the alchemist's delight,
A marvel of nature, shining bright.

Born from the depths of Earth's embrace,
A gift from the heavens, in fiery grace.
Its hue, a blend of fiery red and gold,
A sight to behold, a story untold.

With strength and resilience, it stands firm,
A conductor of energy, ready to affirm.
Its electrons dance, a lively ballet,
Whispering secrets, in their vibrant display.

From ancient civilizations to modern days,
Copper weaves its presence in countless ways.
Adorning statues, bridges, and domes,
A testament to its beauty, as time roams.

But beyond its allure, a deeper tale lies,
Of healing powers, a soothing guise.
For copper, they say, brings balance and peace,
A balm for the soul, a sweet release.

A metal of paradox, both bold and kind,
Copper, a treasure, for all to find.
In its embrace, a world of wonders unfold,
An element of grace, forever untold.

TEN

GREENISH HUES

In the depths of the earth, a treasure lies,
A metal of warmth, a hue that mesmerizes.
Copper, oh copper, your allure so rare,
With tales of ancient civilizations, you share.

A conductor of energy, a link in the chain,
You carry the current, without complaint.
From wires to circuits, you weave your spell,
Guiding electrons, bidding them farewell.

In statues and sculptures, your beauty unfolds,
A testament to craftsmanship, untold.
From the Statue of Liberty, standing tall and free,
To the humble jewelry adorning you and me.

Your presence, so vital, in the realm of life,
A trace mineral, a companion in strife.

From enzymes in cells to the blood in our veins,
You lend your hand, relieving our pains.
　　Yet, there's another side, a darker embrace,
When exposed to the elements, you change your face.
With time, you tarnish, forming greenish hues,
A testament to life's wear and tear, its dues.
　　Copper, oh copper, a metal of grace,
In your essence, a story we can trace.
From ancient times to the present day,
You remain steadfast, in your own unique way.
　　So, let us cherish you, in all your glory,
A symbol of strength, a part of our story.
Copper, oh copper, forever you'll shine,
A chemical element, eternally divine.

ELEVEN

METAL THAT SPARKLES

In a realm where alchemy thrives,
A metal with a coppery hue arrives.
A touch of wonder, a gleam of light,
A story that unfolds, shining bright.

Born from the earth's deep embrace,
Copper, in its raw form, leaves no trace.
Its beauty hidden within its core,
A secret waiting to be explored.

A conductor of energy, it whispers and sings,
With power in its veins, it spreads its wings.
Through wires and cables, it carries the flow,
Connecting our world, from high to low.

In ancient times, a symbol of wealth,
Crafted into vessels, adorned with stealth.

In statues and coins, its value defined,
A precious metal, forever enshrined.

 But copper is more than its monetary worth,
It weaves tales of passion, of love's rebirth.
For when two hearts entwine, like copper they mold,
Strong and enduring, a love to behold.

 Its presence in nature, a dazzling sight,
From fiery sunsets to moon's gentle light.
A patina of green, a weathered embrace,
A testament to time, with elegance and grace.

 Copper, a muse for artists bold,
With brushstrokes of warmth, stories untold.
A canvas alive, with hues so deep,
A masterpiece of copper's eternal sweep.

 So let us raise a toast, to copper's might,
A metal that sparkles, both day and night.
In science and art, it leaves its mark,
A symbol of resilience, shining in the dark.

TWELVE

ADORNED THE CROWN

In the depths of Earth, where secrets lie,
There dwells a metal, both bold and shy.
Its name is Copper, a treasure untold,
A tale of beauty and stories of old.

A metal that gleams with a fiery hue,
Copper dances in the sunlight's view.
Its lustrous glow, a sight to behold,
A symphony of warmth, both new and old.

In ancient times, it adorned the crown,
Of kings and queens, of high renown.
A symbol of wealth, a mark of pride,
Copper stood tall, by their side.

From statues grand to cathedral domes,
Copper weaves tales in ornate homes.

Its gentle touch, a craftsman's delight,
Transforming dreams into works of light.

A conductor of energy, it does embrace,
Electric currents through its veins race.
Powerful and swift, like a river's flow,
Copper conducts, with a radiant glow.

But Copper's true power lies within,
The healing touch it brings, akin.
Its warmth and balance, a soothing balm,
Restoring harmony, bringing calm.

Oh, Copper, a metal so divine,
A legacy that will forever shine.
In tales and dreams, you shall forever remain,
A symbol of strength, love, and gain.

THIRTEEN

REDDISH HUE

In the darkest depths of earth's embrace,
Where molten fires dance with grace,
There lies a treasure, pure and bold,
A metal that shines like burnished gold.

Copper, elemental in its might,
A conductor of both heat and light,
With valiant strength, it weaves its way,
Through nature's hand, in rich array.

From ancient times, it was revered,
By artisans and alchemists, revered,
Crafted into tools, both sharp and strong,
A metal that sings its timeless song.

In verdant fields, it finds its place,
In nature's tapestry, with gentle grace,
Adorning leaves with a copper hue,
A touch of warmth, a vibrant view.

In copper's soul, a story lies,
Of passion, courage, and compromise,
A metal that molds with tender care,
Into sculptures, gems, and treasures rare.

Oh, copper, with your reddish hue,
Your allure is timeless, forever true,
A symbol of love and beauty's reign,
In every heartbeat, in every vein.

So let us cherish this metal dear,
With its lustrous glow, so crystal clear,
For in its essence, we find our worth,
A testament to the beauty of the earth.

FOURTEEN

WARMTH AND GRACE

In the depths of the earth, where minerals reside,
Lies a treasure of old, where secrets reside.
A hue of crimson, a gleam in the dark,
Copper, the element, leaves its mark.

Born in the flames, where heat intertwines,
Its atoms dance freely, like molten wine.
A conductor of energy, both fierce and bold,
Copper's allure, a story yet untold.

From ancient times to modern days,
It weaves through history in countless ways.
Adorning the Pharaohs with a regal sheen,
Or ringing the bells, a melody unseen.

In wires it flows, conducting with grace,
Bringing power and light to every place.

A metal so malleable, it bends to the will,
Crafted by artisans, skilled and still.
 A penny's worth, a symbol of wealth,
Yet copper's true value lies in its stealth.
A healer of ailments, it mends and restores,
A potion of life, a remedy for sores.
 In verdigris patina, a weathered embrace,
Copper's aging beauty, a time-worn face.
With age it evolves, a story it tells,
Of life's many journeys, where wisdom dwells.
 Oh copper, element of warmth and grace,
Your presence brings strength to every space.
In your embrace, a tale to unfold,
Copper, the element, forever bold.

FIFTEEN

FORGED IN THE CORE

In the heart of the earth, where secrets are held,
Lies a metal so precious, so brilliantly melded.
Copper, they call it, with its amber hue,
A shimmering testament, both old and new.

Born from the stars, forged in the core,
Copper emerged, a treasure to adore.
A conductor of energy, a bridge between worlds,
Its magic unfurls as the story unfurls.

From ancient civilizations to modern days,
Copper has played a part in countless ways.
It weaves through wires, carrying power's might,
Illuminating cities in the darkest of night.

In the hands of artisans, copper comes alive,
Crafted into beauty, where art and skill thrive.

As vessels and jewelry, it adorns with grace,
A symbol of elegance, in every place.
    Yet copper bears a secret, hidden from sight,
Its healing properties, a gift of light.
From its touch, ailments find relief,
Restoring balance, mending belief.
    So let us celebrate this elemental wonder,
Copper, a guardian, a protector, and a thunder.
Embrace its warmth, its strength, its art,
For copper's allure will forever impart.
    Let us cherish this element, so bold and rare,
In its essence, a story worth our care.
For copper, like life, is a tapestry to behold,
A testament of beauty, never to be sold.

SIXTEEN

PENNIES THAT GLIMMER

In the heart of the earth, a treasure lies,
A metal that gleams under azure skies.
Copper, a marvel with a fiery hue,
A dance of atoms, an alchemical brew.

    Born in the depths of volcanic embrace,
Forged in the fires, a molten grace.
From ancient times, its secrets unfold,
An element of beauty, a story untold.

    With strength and conductivity, it weaves,
A conductor of energy, the world perceives.
From wires that hum with electric might,
To pennies that glimmer in the pale moonlight.

    In ancient legends and tales of old,
Copper weaves its magic, a story foretold.

A metal of healing, whispered in lore,
With powers to soothe and to restore.
    From the mighty Statue that stands so proud,
To the rooftops gleaming, drawing a crowd.
Copper's touch, a legacy of time,
A testament to its endurance, sublime.
    Oh, copper, thou art a gift from above,
A metal of wonder, a symbol of love.
In your radiant presence, we find our bliss,
A substance of marvel, a metal we can't dismiss.

SEVENTEEN

AGES PAST

In the depths of earth's embrace, Copper gleams,
A lustrous metal, born of ancient dreams.
Its fiery hues ignite the forge of time,
A tale of strength and beauty, so sublime.

Mined from the heart of nature's hidden core,
Copper whispers secrets none have heard before.
A conductor of energy, both fierce and bold,
It weaves a tale of stories yet untold.

In wires and circuits, it pulses with might,
Connecting worlds, bridging day and night.
With every heartbeat of technology's might,
Copper dances in the symphony of light.

Through ages past, it adorned the hands of kings,
A testament of wealth and precious things.

Yet copper's worth lies not in gold or gems,
But in the life it breathes into earth's stems.
    For in the soil, it nurtures life's grand stage,
A catalyst for growth, a guardian of sage.
From vineyards to orchards, fields ripe and vast,
Copper's touch ensures life's harvest will last.
    Oh, Copper, element of strength and grace,
Your presence leaves a mark no one can erase.
A symbol of resilience, forever to endure,
In the tapestry of nature, forever pure.

EIGHTEEN

EARTHLY SPACE

In a world of metals, gleaming bright,
One shines with a captivating light.
Copper, the element of warmth and grace,
A tale I weave, in this sacred space.

From ancient mines, deep underground,
Copper emerges, a treasure found.
Its reddish hue, like a fiery ember,
Reflects the sun in November.

Through the ages, its worth untold,
Crafted into objects, precious and bold.
A conductor of energy, both fierce and pure,
Copper conducts life, that's for sure.

In wires and cables, it carries the spark,
Connecting us all, even in the dark.

From city streets to rural lands,
Copper's touch, a guiding hand.

In statues and sculptures, it takes form,
Captured beauty, defying the norm.
A silent witness, through centuries gone,
Copper's endurance, forever strong.

In medicine's realm, its secrets unfold,
Copper's healing touch, a story untold.
Antimicrobial power, it does possess,
Fighting off germs, leaving no distress.

Oh, Copper, your legacy profound,
In every corner, your presence surrounds.
A symbol of strength, beauty, and lore,
Forever captivating, forever more.

Let us honor your enduring grace,
Oh, Copper, a marvel of this earthly space.

NINETEEN

CRAFTSMANSHIP'S EMBRACE

In the depths of Earth's embrace, a treasure lies,
A metal born of ancient skies.
Copper, the alchemist's delight,
With fiery hues, it shines so bright.

From verdant fields to towering spires,
In every form, its beauty inspires.
A conductor of electric might,
It dances with electrons, pure and white.

Mined from the earth with toil and sweat,
Copper weaves a tale, a secret yet.
Its essence, a gift of nature's art,
A symbol of connection, love at heart.

In wires it carries energy's flow,
From homes to cities, it helps us grow.

In statues and ornaments, it adorns with grace,
A testament to craftsmanship's embrace.
 A penny's worth, but wealth untold,
Copper's value more than can be sold.
Its alchemy lingers in the air,
An element so rare, beyond compare.
 Oh, Copper, guardian of the flame,
With your presence, life will never be the same.
In your embrace, we find solace and might,
A metal so precious, forever shining bright.

TWENTY

SILENT HERO

In the depths of Earth's ancient core,
Where fire and pressure forever roar,
Lies a treasure gleaming bright,
A metal born of nature's might.

Copper, oh noble element bold,
With hues of reddish-gold untold,
Your lustrous beauty, a sight to behold,
A tale of history, yet to be unfold.

From ancient civilizations of old,
To modern wonders, a story untold,
You conduct the currents of life's flow,
A conductor of energy, a radiant glow.

In the hands of craftsmen, you transform,
From sculptures grand, to jewelry adorned,

With malleable grace, you take new form,
A testament to human skill reborn.

    Through wires and circuits, you carry the spark,
Connecting the world, from dawn to dark,
In every device, you play your part,
A silent hero, connecting heart to heart.

    Copper, oh element of strength and might,
Of conductivity, a brilliant light,
In your embrace, the world is bound,
A testament to the wonders profound.

    So, let us celebrate your legacy,
Copper, the element of harmony,
For in your presence, we find our worth,
A symbol of unity, upon this Earth.

TWENTY-ONE

SIGHT TO VIEW

In the depths of Earth, where wonders lie,
A shimmering treasure catches the eye.
A metal so noble, with reddish hue,
Copper, the element, both old and new.

From ancient times, it has been adored,
For its strength and beauty, a metal restored.
Crafted by hands, transformed with care,
Copper gleams, a treasure to share.

In wires and cables, it conducts the flow,
Delivering power wherever we go.
Its conductivity, a gift so rare,
Bringing energy through the air.

In architecture, it stands so bold,
Roofs and domes, a sight to behold.

Weathered and aged, it tells a tale,
Of strength and endurance, it will never fail.

In art and decor, it finds its place,
Adorning spaces with elegance and grace.
Statues and vessels, shining with pride,
Copper's allure, impossible to hide.

From ancient civilizations to modern days,
Copper's legacy forever stays.
A symbol of resilience, a metal divine,
Copper, a treasure of nature's design.

So let us cherish this element true,
With its lustrous glow, a sight to view.
Copper, the metal that captures our heart,
A masterpiece of nature's art.

TWENTY-TWO

COPPER IS VERSATILE

In the earth's deep belly, under layers of stone
Lies a metal that glints in a hue all its own
Copper, they call it, a treasure to behold
Intricate and malleable, worth more than gold
    It's the conductor of energy, sparking and bright
A catalyst for reactions, a chemical delight
It's a part of our history, a symbol of time
From ancient artifacts to modern designs
    Copper is versatile, in many forms it appears
From wires to coins, it's been used for years
Its beauty is unmatched, a shimmering sight
Reflecting light like the stars at night
    It's been a faithful companion to mankind
A metal that's steadfast, always on our mind

From architecture to medicine, it plays a role
A metal that's precious, with a heart of gold
    So let us cherish this element, so rare and fine
For it's an essential part of our design
A metal that's timeless, a treasure to behold
Copper, we salute you, in your glimmering gold.

TWENTY-THREE

WONDER AND GRACE

In the realm of elements, behold Copper's gleam,
A metal of beauty, like a radiant dream.
Its hue, a fiery red, ignites the eye,
A tapestry of warmth, as the flames dance high.

From ancient times, it has graced our sight,
A vessel for artistry, a creator's delight.
From statues grand to intricate designs,
Copper's allure, forever it shines.

Born from Earth's depths, a treasure untold,
A metamorphosis, as time unfolds.
In nature's alchemy, it transforms its state,
From ore to metal, a metamorphosis innate.

A conductor of energy, both vibrant and pure,
In circuits and wires, it ensures power's allure.

With strength and conductivity, it takes its role,
A silent hero, connecting heart and soul.
    Yet Copper's story, not just in its use,
But in its history, a tale of abuse.
For in its extraction, the Earth bears scars,
A reminder to cherish, to heal its wounds with stars.
    Oh, Copper, element of wonder and grace,
May we honor your presence, in every place.
In art and science, in beauty and might,
Copper, forever, you shine in our sight.

TWENTY-FOUR

ANCIENT EGYPT

In the depths of Earth's ancient core,
Where fiery rivers cease to roar,
Lies a metal, both bold and bright,
A treasure kissed by celestial light.

Copper, oh copper, with hues divine,
A lustrous jewel that doth entwine,
With tales of conquest and ancient lore,
A symbol of strength forevermore.

From the hands of craftsmen, copper gleams,
Forged into wonders, a poet's dreams,
A conductor of energy and power,
It weaves connections, hour by hour.

In verdigris patina, a story unfolds,
Of weathered statues, so brave and bold,

Time's gentle touch, a tender embrace,
Each mark and blemish, a sign of grace.
    In alchemy's dance, copper transforms,
From liquid rivers to solid forms,
A catalyst of change, it takes its role,
In nature's laboratory, a vital soul.
    From ancient Egypt to modern day,
Copper's allure will never sway,
A testament to resilience and might,
A metal that shines, both day and night.
    So raise a toast to copper's might,
This element that graces our sight,
A testament to beauty and grace,
Copper, forever, in our embrace.

TWENTY-FIVE

PRECIOUS GIFT

Copper, oh copper, a metal so bright,
A conductor of heat, a conductor of light.
In wires and pipes, you flow with ease,
A crucial part of our modern industries.

From ancient times, you've been revered,
Your beauty and strength, forever endeared.
The Statue of Liberty, a symbol so grand,
Stands tall and proud, holding your hand.

In jewelry and coins, you shine so bold,
A story of wealth and history untold.
Your malleability, a prized attribute,
A sculptor's dream, forever acute.

Your role in our health, often overlooked,
In enzymes and proteins, your presence hooked.

In the blood, you aid in oxygen transport,
A vital element, a medical cohort.
   Copper, oh copper, a metal so divine,
A friend to mankind, a treasure so fine.
From the Earth, you've been bestowed,
A precious gift, forever we'll hold.

TWENTY-SIX

UNIQUE AND RIFE

In the depths of the earth, it lies
A metal that glimmers and catches the eye
A conductor of heat and electricity
Copper is its name, a symbol of nobility

From ancient times, it's been known
For its beauty and strength, it has shown
Adorning kings and queens, it's been used
Its value and versatility, never refused

In the fields of medicine and technology
Copper plays a vital role, a necessity
From wires to pipes, it's indispensable
A metal that's reliable and incredible

But beyond its practical uses, it's more
Copper has a warmth that we adore

A color that's rich and full of life
A metal that's unique and rife
    With history and beauty intertwined
Copper is a treasure that we'll always find
A symbol of strength and endurance
A metal that inspires with its radiance.

TWENTY-SEVEN

COPPER, A MUSE

In copper's gleaming depths, a tale unfolds,
A story of strength, of secrets yet untold.
A metal born of Earth's eternal fire,
With hues that dance, a gift to inspire.

From ancient times, its allure did shine,
A treasure sought, its worth so divine.
Adornments crafted, with artistry and grace,
Copper, a muse, in every time and place.

In alchemy's grasp, it weaves its spell,
A conductor of energy, where wonders dwell.
Electric currents flow, a vibrant stream,
Copper's embrace, a conductor's dream.

In wires and circuits, it finds its role,
Connecting the world, from pole to pole.

A conduit of power, a lifeline unseen,
Copper's embrace, a link in the machine.

Beneath the surface, where minerals lie,
Copper's bounty awaits, a sight to defy.
From mines deep within the Earth's embrace,
Copper emerges, a treasure to chase.

But beyond its strength, its practical might,
Copper holds stories, hidden from sight.
In tales of love, it whispers its song,
A symbol of passion, enduring and strong.

So, let us celebrate this element rare,
With its lustrous glow and secrets to share.
Copper, a marvel, so ancient and new,
Forever entwined, in our hearts it will brew.

TWENTY-EIGHT

METAL DIVINE

In the depths of the earth, a treasure lies,
A metal that gleams with a fiery guise.
Copper, the element, so ancient and bold,
A tale of resilience that's yet to be told.

Its reddish hue, like a sunset's embrace,
Reflects the stories of time and space.
Born in the heart of a blazing star,
It journeyed through galaxies, traveling far.

Mined from the Earth, through labor and sweat,
Copper emerges, a prize to be met.
A conductor of energy, both fierce and true,
It weaves through circuits, bringing life anew.

In ancient times, a gift from the gods,
Forged into tools, adorned by the odds.
A symbol of wealth, a currency of trade,
Copper's allure, never to fade.

From towering statues to humble abodes,
Copper adorns, casting shimmering odes.
As rooftops weather and patinas form,
The beauty of copper is ever warm.

But beyond its luster, there's more to explore,
In a chemist's hands, it reveals even more.
Reactions and compounds, a scientist's delight,
Copper's mysteries, shining ever bright.

So let us raise a toast to this metal divine,
Copper, so noble, enduring through time.
A symbol of strength, both ancient and new,
In its embrace, may we find courage too.

TWENTY-NINE

BREAKING RULES

Copper, oh copper, how you shine so bright
A metal so beautiful, a precious sight
You conduct electricity, you conduct heat
A metal so useful, so incredibly neat
    You turn green with age, a patina so fine
A sign of character, a mark of time
From coins to wires, you serve us all
A metal so versatile, standing tall
    You were once used for weapons and tools
But now you're in technology, breaking rules
Phones, computers, and cars too
A metal so important, it's true
    From ancient times to modern day
You've been a part of our lives in every way

Copper, oh copper, how you shine so bright
A metal so beautiful, a precious sight
   So here's to you, copper, a metal so rare
A symbol of strength, a metal so fair
May you continue to shine and never fade
A metal so valuable, never to evade.

THIRTY

STRENGTH AND DURABILITY

In the depths of earth's embrace, Copper resides,
A metal of warmth, where secrets reside.
Its hue, a burnished beauty, reddish-brown,
Reflects the tales of old, yet to be known.

From ancient mines, it emerges with grace,
A treasure sought by many in this vast space.
A conductor of energy, both fierce and strong,
Copper's essence dances with a melodious song.

Its touch, a healer, soothing to the skin,
A symbol of love, a connection within.
Adorned in copper, we forge a sacred bond,
A bridge between hearts, forever beyond.

In alchemy's realm, it transforms with fire,
Unveiling its true form, a gleaming desire.

With strength and durability, it stands the test,
Resilient and timeless, unlike the rest.

In tales of old, artisans weave its story,
Crafting masterpieces, an ode to its glory.
From towering roofs to stately statues grand,
Copper's presence, a touch of the artist's hand.

Oh, Copper, element of splendor and might,
A marvel of nature, shimmering in light.
Through time and space, your legacy endures,
A symbol of beauty, forever pure.

THIRTY-ONE

UNBROKEN AND RAW

In the heart of the earth's deep embrace,
There lies a treasure, full of grace.
A metal revered for its radiant hue,
Copper, a marvel, ancient and true.

Beneath the surface, where secrets reside,
Copper emerges, a gift to confide.
Its lustrous glow, a fiery dance,
Reflecting memories of an ancient romance.

A conductor of energy, both fierce and bold,
Copper weaves stories, never to be told.
Through wires and circuits, it carries the spark,
Igniting the world, even in the dark.

In art and design, it finds a home,
A canvas for craftsmen, where beauty is sown.

From statues to jewelry, its touch is divine,
Crafting masterpieces, for all to entwine.

But beyond its allure, Copper holds more,
A healer of bodies, a remedy to restore.
With antimicrobial powers, it fights the fight,
Defending our health, in the darkest of night.

Oh, Copper, element of wonder and awe,
Your legacy stretches, unbroken and raw.
A symbol of strength, resilience, and grace,
Forever engraved in time and space.

THIRTY-TWO

CONDUCT HEAT AND ELECTRICITY

Copper, oh copper, you shine so bright
A metal that's pure, a hue so right
Your reddish-brown color, a sight to behold
A beauty unmatched, a story untold
In ancient times, you were highly sought
A symbol of wealth, a treasure brought
From mines deep within the earth
You were extracted with great mirth
You conduct heat and electricity with ease
A property that's rare, a gift to seize
In wires, pipes, and coins, you're used
In jewelry, statues, and art, you're fused
Copper, oh copper, your value so high
A metal that's versatile, a reason to buy

You're a part of our daily lives
In the form of electronics and hives
    Your history, your present, and your future bright
Copper, oh copper, you shine so bright

THIRTY-THREE

MAGIC IN THE SAND

In the depths of Earth's embrace, Copper lies,
A metal born of ancient fires in disguise.
Its lustrous hue, a dance of amber and red,
A story of resilience, by time and nature bred.

From mines deep within the earth, it's unearthed,
A treasure sought by those with hearts immersed.
Malleable and strong, it bends to the hand,
Crafted into art, a testament to man's command.

In wires and cables, it conducts the flow,
Electricity's ally, a current's gentle glow.
In the coins that jingle, it carries value and worth,
A symbol of wealth, a currency of the earth.

Yet Copper's tale extends beyond its use,
For it holds secrets, a hidden muse.

It whispers of alchemy and ancient lore,
Of healing properties, a remedy to explore.

For in its essence, Copper holds a trace,
Of energy and balance, a healing embrace.
From arthritis to fatigue, it lends a helping hand,
Restoring harmony, a touch of magic in the sand.

So let us honor Copper, this element divine,
A gift from nature, a treasure we intertwine.
In its beauty and wisdom, may we always find,
The strength to endure, with a heart refined.

THIRTY-FOUR

IGNITES THE DARK

In a realm of alchemy's embrace,
Where elements entwined, grace,
There lies a metal, copper by name,
With secrets in its fiery flame.

Born from the depths of Earth's core,
A treasure coveted, forevermore,
From mines deep, its veins do run,
A gleaming treasure, second to none.

A conductor of energy, grand and true,
Binding currents, flowing through,
In wires and cables, it finds its might,
An electric symphony, shining bright.

Its hue, a ruddy shade of red,
A testament to the fire it's fed,
Transforming, evolving, as time goes by,
From copper's grip, no secrets lie.

In ancient times, it took its place,
Adorning temples with its grace,
A symbol of wealth, prosperity,
A metal that shaped history.

From humble beginnings, it did ascend,
Transformed by artisans, who would mend,
Creating vessels, ornate and bold,
With copper's touch, stories unfold.

Oh, copper, you are a metal divine,
Your legacy, forever will shine,
A testament to human endeavor,
In your presence, we find forever.

So let us celebrate your copper hue,
A metal that brings both old and new,
In art and industry, you leave your mark,
Copper, the flame that ignites the dark.

THIRTY-FIVE

METAL OF MIGHT

In depths of Earth's forgotten mines,
Where darkness reigns and silence pines,
There lies a treasure, rich and bold,
A metal with a story yet untold.

    Copper, the element of ancient lore,
Gleaming with a reddish-orange core.
Its beauty, a tapestry of russet hues,
A testament to time, it never loses.

    From the dawn of civilization's birth,
Copper was cherished for its priceless worth.
A conductor of energy, both fierce and free,
Binding the world with its electrical spree.

    In alchemy's mystic dance it thrived,
Turning base metals, dreams revived.

An alchemist's secret, a clandestine brew,
Copper's transformation, its magic grew.

From delicate wires to towering spires,
Copper weaves through our electric desires.
In art and architecture, it finds its place,
A testament to its enduring grace.

But let us not forget its humble past,
When copper coins were meant to last.
A token of wealth, a symbol of might,
Copper's legacy, shining bright.

And as the world turns, its story unfolds,
Copper's allure, forever beholds.
A mineral of wonder, a metal of might,
Copper's presence, a guiding light.

ABOUT THE AUTHOR

Walter the Educator is one of the pseudonyms for Walter Anderson. Formally educated in Chemistry, Business, and Education, he is an educator, an author, a diverse entrepreneur, and he is the son of a disabled war veteran. "Walter the Educator" shares his time between educating and creating. He holds interests and owns several creative projects that entertain, enlighten, enhance, and educate, hoping to inspire and motivate you.

Follow, find new works, and stay up to date
with Walter the Educator™
at WaltertheEducator.com

www.ingramcontent.com/pod-product-compliance
Lightning Source LLC
LaVergne TN
LVHW051959060526
838201LV00059B/3730